The Rise of Allopathic Medicine

Junk Science?

Ryder Management Inc.

Copyright © 2015 Ryder Management Inc.

All rights reserved.

ISBN-10: 1505896096
ISBN-13: 978-1505896091

Prologue

"If people let the government decide what foods they eat and what medicines they take, their bodies will soon be in as sorry as state as are the souls who live under tyranny."

By Thomas Jefferson

Table of Contents

Prologue	iii
Table of Contents	v
Introduction	1
Homeopathy vs. Allopathic Medicine	2
Medicine as Science	3
The Flexner Report	4
American Medical Association (AMA)	5
Food and Drug Administration (FDA)	6
The Rockefeller Foundation	7
United Nations (UN)	8
Junk Science	9
Standard Oil	11
Natural Remedies	12
Health Tips	14
Conclusion	15
ABOUT THE AUTHOR	16

Introduction

Did you know that in the early part of the 1990's, the **cause** of many diseases, including autoimmune diseases, was found to be a deficiency in micronutrients? Diseases that are caused by a micronutrient deficiency include, but not limited to high blood pressure, heart attacks, cancer, AIDS, Type 2 Diabetes and most other autoimmune diseases. In addition, this finding has been deliberately kept from the masses for reasons that amount to greed. Who suppressed the findings? It was the pharmaceutical industry, also known as "Big Pharma".

The purpose of this eBook is to provide you with information on the history of medicine and the pharmaceutical industry. After you have finished reading this eBook, hopefully you will question the efficacy of the health care you receive.

Ryder Management Inc.

Homeopathy vs. Allopathic Medicine

In the 1800's, patients had a choice between using the services offered by either Homeopathic Doctors or Allopathic Doctors.

The allopathic doctors believed that physicians must aggressively drive disease from the human body, basing their practice on what they considered scientific theory. The Allopath used three main techniques: they bled the body to drain out the bad humors; they gave huge amounts of toxic minerals such as mercury and lead to displace the original disease; and they also used the brutal procedure of surgery in an attempt to cut out the disease from the human body. It's no secret why most people feared allopathic doctors altogether. During this time in the late eighteen hundreds, the perception amongst the masses was that patients died of the cure for the disease.

Contrary to the allopathic doctors, the Empirics or Homeopathic Doctors believed in stimulating the body's own defense mechanisms to heal itself from disease. Instead of poisonous minerals, the Empirics used natural products such as herbs and other natural plants learned from Native Americans and old European traditions. Their methods were not based on theories, rather, they were based on observations and experience. With this method, patients who did pass on, passed on as a result of the disease, not the cure. Homeopathic medicine is nature based and is available in abundance at low cost.

It wasn't until the turn of the twentieth century when new profitable procedures were discovered that allopathic doctors secured a firm hold on the industry.

Medicine as Science

Changes in American medical practices from 1870 until the outbreak of WW 1 in 1914, reflected the gradual acceptance of recent developments in Europe, particularly in Germany and Austria where medical research was just beginning.

The fortunes of Carnegie, Morgan and Rockefeller provided the foundation of the new emerging allopathic medical economy. Ironically, John D. Rockefeller, who invested heavily in the emerging allopathic medical industry, insisted on using only homeopathic remedies for his own health.

Rockefeller wanted total domination and control over not only the petroleum and petrochemical industries, but also the new emerging pharmaceutical industry, whose products are derived from "coal tars" or crude oil. By using his corporation "Standard Oil of New Jersey", he obtained a controlling interest in the large German drug cartel called "I.G. Farben", the fourth largest industrial concern, behind General Motors, U.S. Steel and Standard Oil of New Jersey. It should also be pointed out that IG Farben was notorious for its role in the Holocaust.

The campaign to reduce and eliminate competition from non-drug therapies began with the acceptance of the Flexner Report in 1910.

The Flexner Report

Abraham Flexner was engaged by J.D. Rockefeller to evaluate the effectiveness of "institutions of the healing arts". The ultimate report was submitted to the Carnegie Foundation and was entitled *"Medical Education in the United States and Canada"*. Page 22 of the report stated "the privileges of the medical school can no longer be open to casual strollers from the highway. It is necessary to install a *doorkeeper* who will, by critical scrutiny, ascertain the fitness of the applicant, a necessity suggested, in the first place, but consideration for the candidate, whose time and talents will serve him better in some other vocation, if he be unfit for this, and in the second, by consideration for a public entitled to protection from those whom the very boldness of modern medical strategy equips with instruments that, tremendously effective for good when rightly used, are all the more terrible for harm if ignorantly or incompetently employed."

Following the submission of the Flexner report, it was decided that the American Medical Association (AMA) would become the *doorkeeper*. The AMA then became empowered to certify or decertify any medical school in the country on the basis of whether a school met the standards of "approved" medicine established by this organization.

In November of 2013, the AMA reaffirmed its opposition to cannabis legalization. The AMA's 527 member House of Delegates decided to retain its long standing position that "cannabis is a dangerous drug and as such is a public health concern" despite proof of its unique disease curing ability.

.

American Medical Association (AMA)

The American Medical Association (AMA), since its inception in 1847, was hostile to the interests of certain proprietary medical colleges and their faculties. In 1874, the AMA abolished medical colleges' voting rights in the association; thus establishing control of the medical discipline. The AMA operated by a "code of ethics" which was used to deny a patient the right to judge their physician. The code was also used to encourage attacks on "irregular doctors" who were also dubbed "quacks". The AMA's goal was to reduce and ultimately eliminate any and all types of perceived competition with pharmaceutical based medicine. At the same time that the AMA was adamant with increasing the low standards that existed for medical education, they also demanded a complete trust of doctors by their patients.

Today, the AMA administers a medical coding system in which every health-care professional and hospital in the US must use in order to be paid for services rendered. This monopoly generates an annual income of approximately $70 to $100 million for the AMA.

Food and Drug Administration (FDA)

The Food and Drug Administration (FDA) was formerly known as the "Division of Chemistry" prior to being called the "Bureau of Chemistry". However, Rockefeller wanted to hide its connection to the chemical industry agenda and changed its name in 1913 to the Food and Drug Administration. 1913 was also the same year that the Rockefeller Foundation and the Federal Reserve were founded.

The FDA's role is to approve and sanction products that they deem as generally recognized as safe (GRAS).

Due to the FDA's connection to the chemical industry, the FDA favored those companies that put chemicals into their food or medical products during its approval process. This favoritism includes the pharmaceutical industry's chemical laden pills. This favoritism has been referred to as providing the chemical industry with "legal immunity" since people do not have legal recourse against the chemical industry. The FDA shields the chemical industry from liability by providing them with "plausible deniability".

Contamination issues are withheld from the public by routinely classifying such issues as "P.R. related".

The FDA is also used as a means for eliminating competition to the multi trillion dollar pharmaceutical drug industry.

The FDA works closely with the Rockefeller Foundation, the AMA and the chemical industries

The Rockefeller Foundation

The Rockefeller Foundation (RF) was formed in 1913 and its main emphasis was on medicine and medical education. The first Rockefeller medical school donations amounted to $550 million dollars. In 1928, the Rockefeller Foundation advanced money to eighteen medical schools in fourteen countries. It was during this time that an aggressive campaign started by the FDA to suppress any medicine that competed with the chemical industry.

On January 18, 1982, Dr. Richard J. Crout, the Director of the FDA's Bureau of Drugs, was quoted as follows:

"I never have and never will approve a new drug to an individual, but only to a large pharmaceutical company with unlimited finances."

In Montreal, Canada, the McGill University Press reported the following:

"In 1919, a five million dollar Rockefeller Foundation gift to certain Canadian medical schools helped bring Canadian medical education to the 20th century."

Medical schools who became "compliant" received Rockefeller grants, since *"compliance"* was a condition of the money being handed out by the Rockefeller Foundation (RF). "Compliance" required abandoning traditional and natural forms of medicines in favor of a new generation of using petrochemicals for drugs. This also included trashing entire libraries of historical procedures.

To ensure compliance from the medical schools in receipt of RF grant money, Rockefeller employees were placed on the medical school's board of directors.

United Nations (UN)

In order for Rockefeller's oil and drug business to thrive, it was necessary to create the right political framework - enter the United Nations (UN).

The United Nations (UN) was created in San Francisco in 1945 and to seize control of the post war era medical industry and the three leading drug export nations had all the say while the 200 other nations were rendered mere spectators.

The United Nations was founded as an organization to allegedly serve the well-being of people all over the world. Through the UN's subsidiary organizations including the World Health Organization (WHO) and the World Trade Organization (WTO), all three organizations are actually nothing more than the political arms that serve the oil and drug industries at the global level.

In 1963, under the pretense of consumer protection, the United Nations launched a four decade long mission of outlawing vitamin therapy and any other natural and unpatentable approach to healthcare within all member countries of the United Nations. The goal was to ban all competition for the then multi-billion dollar pharmaceutical industry and their patented pills.

Junk Science

Junk Science is a term that refers to faulty scientific research, data analysis along with the claims created for financial or political gain. The concept is often brought up or invoked in political and legal contexts where scientific information has a great deal of weight in decision making. The term originates from a tort legal case in the mid 1980's.

This term is introduced here because the above definition describes perfectly the activities of the pharmaceutical industry (Big Pharma).

Following is a partial list of how Big Pharma has used the media to sell us their junk science:

Using deceptively low doses in scientific testing: It has been discovered that behind the published reports that state vitamin supplements have no effect on health actually obtained that result by using dosages at very low levels. In other words, in order to conspire to have scientific evidence report desired results, the nondisclosure of the amount used to affect the result was hidden and not reported.

Biased media publications: publications are actually owned by Big Pharma, Big Food or Big Ag. In addition, most drug trials are not made available to the public and negative findings are hidden.

Parroting press releases which actually come from one of the Big Three (Pharma, Food, Ag) marketing departments rather than independent research and analysis.

Cherry picking conclusions and only reporting desired results.

Ghostwritten studies by Big Pharma but published under a physician's name.

Questionable methodologies

Skewed analysis

In terms of chemical imbalances and psychiatry, according to Dr. Thomas Szasz, no one to date, has developed a test, measurement or demonstrated the science of "chemical imbalance" and as such, there can be no such thing.

Standard Oil

Instead of just owning and controlling all petroleum, the Rockefeller Empire, by the middle of the twentieth century, controlled virtually all medicine and medical education. They also owned and controlled the majority of the chemical industry which would eventually become known as the "pharmaceutical industry".

As previously mentioned, John D. Rockefeller himself only used traditional holistic medicines for his own health.

After the kidnapping and taking over of the medical industry, the true carnage from diseases began which included polio, heart disease and cancer.

Also noteworthy, the American Cancer Society was also founded in 1913 by John D. Rockefeller, Jr.

According to the US Census, the US has the most dangerous yet most expensive medical system in the world and it causes 60% of all US bankruptcies.

Natural Remedies

All plants are naturally made of phytochemicals and have the ability of eliciting specific physiological and energetic effects in our bodies. Each plant contains a unique balance of phytochemicals and their specific therapeutic qualities can be used as specific healing agents.

Following is a partial list of commonplace diseases along with plants that can be effectively used in reversing specific diseases. This list is by no means complete as there are a plethora of natural remedies just as there does exist a multitude of plant species.

Liver: Three of the top liver disease reversing herbs includes milk thistle, dandelion and burdock. All three of these natural herbs stimulate the production of bile along with protect, restore and produce healthy new liver cells. Another remedy is the regular use of coffee enemas.

Heart: Coenzyme Q10 (CoQ10), spirulina, hemp, Omega-3 fatty acids, Green tea, Pomegranate, Magnesium and Potassium

Lyme disease: Cannabis oil with a low THC content and high CBD.

Cancer: Cannabis oil, laetrile found in apricot seeds, turmeric, baking soda, mistletoe, burdock root, sheep sorrel, slippery elm, coffee enemas; diluted food grade hydrogen peroxide, hemp seed, to name only a few.

Pain: White willow bark, cannabis, turmeric, coffee enemas

Anemia: Hemp seed products - hemp seed is high in iron and B vitamins including B12; spirulina is another superfood loaded in

B12, beta carotene and many other powerful vitamins and nutrients.

Autoimmune: Vitamin D of approximately 5,000 IU's per day; coconut oil; organic oregano oil; hemp seed; juiced cannabis leaf, turmeric, pumpkin seeds

Depression: Turmeric, organic chocolate, St. John's Wort, hemp

High Blood Pressure: Hawthorn, oregano, cardamom, garlic, cinnamon, olives, onions

Diabetes: Cat's claw; cinnamon; black seed oil (Nigella Sativa), bitter gourd, cayenne pepper, prickly pear, iodine, Pau d'arco, Ho Shou Wu

Health Tips

If you are unable to eat something, you should not put it on your skin.

If you are unable to pronounce something, you should not eat or use it on your skin.

Fluoride (sodium fluoride) is BAD for your teeth and health.

Almost all corn, soy, canola, cottonseed oil, gluten products and sugar beets have been genetically modified (GMO). These products contain pesticide and herbicide in their seeds.

Never consume food that was made in a factory. Artificial sweeteners and everything GMO is factory made.

Animals that are used for milk and meat production today have been fed artificial growth hormones which are toxic to their health as well as our health.

Wash your hands with soap and water on a regular basis.

Knowledge is power! Do your homework and research for yourself, what the real truth is. What you don't know, can kill you.

In Closing

This book has given you a brief look back over the last two hundred years and how it has formed medicine as we know it today. In terms of our own individual health and well-being, it is important to do your own homework in order that you can make an intelligent decision over your own healthcare, one that could very well save your own life.

.

ABOUT THE AUTHOR

Ryder Management Inc. (Rydermgt or RMI) is a Canadian Controlled Private Corporation (CCPC) based in London, ON Canada. As an "umbrella" organization, RMI brings together a group of authors whom are professionals in their respective fields and are writing with the primary goal of providing books that educate, comfort and offer assurance that natural health remedies do exist and are an effective and safe way to regain, obtain and maintain our health.

Please see other books written by Ryder Management Inc at their Amazon's author page at the following:

http://www.amazon.com/Ryder-Management-Inc/e/B00ICGMCRS

Sources of References

Natural News: naturalnews.com

The Health Rangers Wiki: truthwiki.com

EnerHealth: wnerhealthbotanicals.com

Ryder Management Inc: RyderManagement.ca

www.ingramcontent.com/pod-product-compliance
Lightning Source LLC
Chambersburg PA
CBHW070737180526
45167CB00004B/1785